Generis
PUBLISHING

Aquaculture:
A Tool for Community Sustainable Development

Under the Background of Cultural and Economic Consideration in Nigeria

Dr. Babagana Zanna

Title: **Aquaculture:**
A Tool for Community Sustainable Development

Under the Background of Cultural and Economic Consideration in Nigeria

ISBN: 979-8-89966-389-5

Author: Dr. Babagana Zanna

Cover image: www.pixabay.com

Publisher: Generis Publishing
Online orders: www.generis-publishing.com
Contact email: info@generis-publishing.com

Aquaculture: A Tool for Community Sustainable Development under the Background of Cultural and Economic Consideration in Nigeria

Dr. Babagana Zanna

Federal College of Freshwater Fisheries Technology, Baga, Maiduguri, Borno State, Nigeria

DEDICATION

I dedicate this research work to my late mother Yakori Kazallaram (Khadija), may her soul rest in peace – Ameen.

ACKNOWLEDGEMENTS

I begin by expressing my gratitude to our Creator, Almighty Allah, for granting me life and good health, enabling me to successfully complete this research work. I would like to express my sincere appreciation to the Centre for the Study and Promotion of Cultural Sustainability, University of Maiduguri, for supporting me to undertake this research as part of the assessment requirements for the award of Doctor of Philosophy (PhD) in Cultural Sustainability.

I am deeply grateful to my supervisors and lecturers; Prof. Musa Muhammad, Department of Economics, Prof. Abba Isa Tijani, Director, Centre for the Study and Promotion of Cultural Sustainability, and Prof. Garba Ibrahim, Department of History, for providing invaluable guidance and support throughout the period of the research and in my academic pursuit. I appreciate the contributions of Prof. Saliba B. James, Director, Centre for Archives, Record Management, and Training and Prof. Christopher Y. Mtaku, Deputy Director, Center for the Study and Promotion of Cultural Sustainability all of the University of Maiduguri, in the attainment of this success. I thank all lecturers and Staff of the Center for their support.

Special thanks to Prof. Mala M. Daura, Director, Center for Conflict Resolution, Disaster and Risk Management, University of Maiduguri, for his unwavering support. I acknowledge Dr. Abba Alkali Goni, Department of Language and Linguistics, University of Maiduguri, for his intellectual and logistical support.

My heartfelt thanks to my family – my wife, Yagana Ali, and my children, Maryam, Umar, Fatima, Abba Waziri, Khadija Ammi, Falmata Mama, Fatima II Baana, and Muhammad – for their supports and active participation in the conduct of the research. I am grateful to the women fishers and community members in the study area for their cooperation and participation.

Finally, I express my gratitude to **His Excellency, Prof. Babagana Umara Zulum, CON, mni, FNSE, Executive Governor, Borno State,** for providing a grant for the research and implementing the outcome through training and empowerment schemes, contributing to poverty reduction and economic growth in the study area and Borno State at large.

TABLE OF CONTENTS

ABSTRACT

The study is an attempt to examines aquaculture as a tool for community sustainable development with emphasis to cultural and economic consideration in 303 housing estate community, “an area located within Sudan-Sahelian zone” Jere Local Government Area, Borno State of Nigeria. Data for the study were collected through face to face interview from the 22nd to 27th of each month May, June and July, 2019. The results showed that majority of the fishers were women and aquaculture production system was profitable in the study area. The fishers prioritized aquaculture production system; it had contributed positively to fishers’ livelihood and enhances sustainable fisheries development. Aquaculture was not in any way contradictory to the principle of women in fishing activities neither culturally nor religiously. Regulatory framework was not in existence and other support system from government and non-governmental organizations was poor. Recommendations were made that the government, non-governmental organizations, cooperative societies and other major stake holders in the system should initiate policies that will promote aquaculture for the attainment of community sustainable fisheries development in the study area and developing countries of the world.

Keywords: Aquaculture, Community, Cultural, Economic, Livelihood, Regulatory framework, Sustainable development, Tool

1. CHAPTER ONE: INTRODUCTION

Passed 7 billion people in 2012, the human population on earth is expected to reach 9 billion people in 2050 (Garcia and Rosenberg, 2010)], with the resultant need for increased nourishment. It is known that the natural marine fish stocks cannot provide the need for proteins of this increase. This is because their biomasses have continued to decline and that in 2011, 61.3% were fully utilized and could no longer be harvested at a biologically sustainable level (FAO, 2014). The primary method of the past several decades has therefore been to expand the aquaculture industry for all farmed species, from freshwater prawns (*Macrobranchium rosenbergerii*) in the global South to Atlantic salmon (Salmo salar) in Norway (Abdalla and Sumaila, 2007). In 2010–2012 numbers, aquaculture accounted for 41% of total global food fish supply. It is also the fastest-growing animal food producing sector globally. The sector is furthermore expected to supply more than 50% of fish for human consumption by 2015, and 53% by 2022 (FAO, 2014). As such, it has the potential to contribute substantially to the global food supply (Cunning, 2010).

Okoye and Ayanda (2004) in the last three decades, aquaculture has recorded a significant and most rapid growth among the food-producing sectors and has developed into a globally robust and vital industry. Aquaculture has the potential for providing significant volumes of fish and other aquatic food for human consumption, creating substantial employment and reducing poverty, often in remote areas. Aquaculture development in most countries of the world started over 50 years ago was believed to bridge the gap between fish demand and supply. Aquaculture has been pointed out to be catalytic to food security, hunger reduction and poverty alleviation through economic growth and employment generation particularly at this period of serious declined in fish production from capture fisheries as a result of exploitation strategies and land use practices that are antithetical to the principles of sustainable development coupled with the state of insecurity in some part of world e.g. Lake Chad Basin Areas of Nigeria, Cameroon, Niger and Chad Republic.

The development of aquaculture in rural communities in most African countries, has been very slow for several reasons; lack of feeds and high seeds (fingerlings), inadequate access to credit, environmental degradation, conflict with other sectors, poor experience of past attempts at developing aquaculture, inadequate and in appropriate research on the aspects of aquaculture and lack of economic viability studies (Okoye and Ayanda, 2004).

Abbas, Damisa, and Ahmed (2015) in Nigeria, annual national fish demand stands at about 27 million metric tons while domestic fish production is about 800,000 metric tons. It is particularly interesting to know that less than 30% of the total annual fish consumed by Nigerians are produced locally. Aquaculture in Nigeria is ignored by policy makers and development professionals even though it has huge prospects in alleviating poverty, malnutrition, serves as a source of foreign exchange earnings and also means of sustainable community development.

Nnodim (2014) Nigeria is endowed with many large rivers, manmade lakes, creeks and about 200 nautical miles of marine water under the exclusive economic zone, however, the performance of the fishery sub-sector is still far below expectation with inadequate domestic fish supply. Domestic fish production in Nigeria stands at about 800,000 million metric tons leaving deficit of 1.9 million metric tons. This is evident to the fact that Nigeria still imports fish in to the country to supplement local fish production. Nigeria spends about N125 billion annually on the importation of fish. The current import dependent situation is deemed to be unfavorable and non-optimal to the Nigerian economy considering the aquaculture potentials of the country. The continuous importation of fish portends a grave danger to Nigeria in term of foreign exchange earnings, its drain on the foreign reserves and the loss of employment opportunities for Nigerian especially the rural community thus aggravating their poverty level more so a major setback to community sustainable fisheries.

Raji and Omoyemi (2005) the harnessing development and proper use of water resources have often served as yard stick for evaluating sustainable development in riparian communities. The aquatic ecosystem is bound to respond to the exploitation strategies and practices. If the practices are deleterious, the impacts are bound to be negative. At the Lake Chad Basin, most exploitation strategies and land use practices are antithetical to the principle of sustainable development such as obnoxious fishing practices e.g. dumba fishing (digging of small space holes), inappropriate agricultural practices, indiscriminate grazing of grassland vegetation, reckless fuel–wood cutting, creation of borrow (harrowing pits) associated with bad mining practices and brick making, water pollution. What are evident at the Lake Shores in the last twelve (12) years were population pressure and the continuous exploitation of marginal lands especially deforestation have continued to aggravate the process of drought and desertification in the area. With the emergence of "Boko Haram" Insurgency, fish supplies from such area have drastically fallen. To meet the challenges of sustainable fisheries development in some fishing communities, it is considered necessary to embark on aquaculture fish farming.

FAO (2006) in Nigeria, the population of fish is about 0.7 million metric tons annually which results in shortfall of about 1.0 million metric tons annually. Only 5% of this 0.7% million metric tons produced locally is from aquaculture. The remaining 95% is from the capture fisheries, which are dominated by the artisanal fishermen. Out of 35 grams of animal protein per day per person recommended by food and agricultural organization, less than 7 grams is consumed on the average. As a result of this, many Nigerians suffer from protein deficiency due to low protein intake.

Inoni (2001) The current low supply trends with ever increasing population combined with the state of in-security in some part of world, the per capital consumption of fish in some fishing communities is stagnating and supply to other part of the country has drastically fallen e.g. Nigeria. The resource utilization practice being undertaking by local fishing communities against the background of community, the nation and the world at large is another problem. The bulk of domestic fish production in Nigeria is from capture fisheries which are not sufficient to satisfy the ever-growing demand for fish. He also observed that there is a decline in yield of natural fish stocks which is an indicator that fish stocks have reached the point of maximum sustainable yield since the number of catch from capture fisheries is declining. Factors implicated in the decline of capture fisheries include climate change, oil spillage and discharge of industrial wastes in to river which results in pollution of river where people fish from, over fishing and rudimentary technology of fishing.

1.1 Statement of the Problem

Aquaculture in contrast with capture fisheries little has been done to improve the scope of fishers' participation in aquaculture or to bring fishers' and new people in to mainstream aquaculture development effort of most communities in Borno state of Nigeria for the attainment of sustainable fisheries. Aquaculture development in other part of the African countries has impacted on social-economic objectives such as nutrition improvement in rural areas, income generation, diversification of farm activities (integrated farming) and creation of employment especially in rural communities where opportunities for aquaculture and its economic activities are limited. This approach over years has resulted in sustained aquaculture growth in some part of African countries i.e., Gabon, Senegal (Stella, 2000). Yet there is still room for enhancing aquaculture production in such areas through improved production systems (cost effective and high-quality fish feed), genetics and general farm management principles to achieve the desired and expected growth of aquaculture to meet the ever -increasing demand for fish and satisfy its socio-economic functions for community sustainable development.

The practice of aquaculture is below expectation in Jere Local Government Area, Borno State of Nigeria in spite its location within an area that met environmental requirement for aquaculture production both in term of water, soil and temperature, regardless of the fact that aquaculture ranking is becoming or has already become the dominant factor in fish supply. To arrest this deplorable condition to boost fish production, aquaculture remains the only feasible option that can sustain adequate fish supply for the development of communities.

1.2 Research Objectives

The main objective of the study is to examine the viability of aquaculture as a tool for sustainable development in fishing communities under the background of economic and cultural factors in Jere Local Government Area Community, Borno State of Nigeria. The specific objectives are to:

1. assess aquaculture at the stand point of economic and cultural consideration for community sustainable development;
2. examine the integration of less intensive aquaculture systems in to community development;
3. assess the promotion of more secure access to resources and better management of the resources at community level;
4. examine existing community environmental and other policies and skills for the exploitation of aquaculture opportunities in the study area;
5. assess domestic fish production and development of local fisheries-based industries in the study area;
6. assess the factors militating aquaculture in community sustainable fisheries development in the study area.

It is hoped that the outcome of this study may provide a basis for an approach to interventions on improving the condition of aquaculture fish farmers in the fishing communities. The information acquired may strengthen the strategy of project and government planners in designing measures in the development of aquaculture fisheries. Researchers may use the research outcome for references and other findings. It may also serve as a source of data to individuals, groups, agencies, organizations both governmental and non–governmental for decision making in order to produce actions that may improve the fishing communities, their living and working conditions for the overall development and sustainability of the fisheries sector of the economy.

The study was carried out in 303 housing estate community of Jere Local Government Area, Borno State of Nigeria. The study focused on available aquaculture fishers in the study area. Interview for the research work was carried out from 22nd to 27th of each month "May, June and July, 2019" in consideration of the end of production period of most farmers associated with intensive fish marketing in the study area to enable the acquisition of accurate and reliable information.

2. CHAPTER TWO: LITERATURE REVIEW

Aquaculture is referred to as the farming of aquatic organism in land and coastal areas, involving intervention in the rearing process to enhance production and the individual or corporate ownership of the stock being cultivated. In rural areas of most developing countries, agronomy, water management, aquaculture and wild aquatic resource harvesting are often physically and functionally integrated. Thus, aquaculture is an integral and indivisible part of the management of aquatic resources (Haylor and Bland, 2001).

FAO (2007) Food Security has been defined as a condition when all people, at all times, have physical and economic access to sufficient, safe and nutritious food to meet their dietary needs and food preferences for an active and healthy life, this concerns not only food production and distribution but also has social, economic and institutional dimensions. More so, Olufayo, (2012) affirmed that participation of aquaculture in Nigeria is increasing daily and this has contributed to house hold food security in the followings:

1. Increased availability of fish market.
2. Help to provides food of high nutritional value especially for vulnerable groups such as pregnant and lactating women, infants and pre-school children.
3. Provides employment.
4. Provides highly digestible proteins that are rich sources of fat and water – soluble vitamins, minerals and fatty acids.
5. Provides a focal point for agricultural diversifications and increased sustainability by providing a source of water.
6. Increase farm sustainability through construction of ponds, which also serve as small–scale, on farm reservoirs.
7. Reduces some social, cultural and economic constraints women are going through in performing their household roles.

Serge (2007) Sustainable development means the management and conservation of the natural resource base and the orientation of technological and institutional change to ensure continued supply of human needs for present and future generations. Sustainable fisheries development is defined as fisheries development that integrates bioecological, economic and social dimensions to sustainably improve the well-being of all the people engaged directly or indirectly in the fisheries sector as well as natural productive system.

Patricia and Julia (2007) characterization of fishing communities ranges from containers of cultural traditions to sites of resistance, from place of management to space of global connection. (Barth, 1992) "the degree of order in social interactions, the degree of sharing of culture, and the adequacy of mutual interpretations are questions that can be studied in local community and represent the effect of processes which can be studied and identified. According to (Fischer, 1975) fishing communities because of their variety, the demands of fishing itself and the heightened awareness that comes from vulnerability after rich insights in to these questions. Ultimately, however, "whether one should assume that communities are significantly variable objects must be determining by the empirical utility of that assumptions, and with regard to specific phenomena, The need for comparative research with specific framework. (Kurlansky, 1997) anthropologists have long realized that communities and culture are rarely isolated from political and economic linkages with other culture and places and that globalization is not new, given centuries of trade, conquest and migration. (Massey,1993) observed that understanding such linkages, however, involves culturally nuanced and historically changing notions of place economy and identity. According to (Lipuma, 1992) in Galica "national and ethnic identities have become attached to primary industries, such as fishing but also farming and wire making. (Rogers, 1991) confirmed that labor in its full capitalist form has not totally replaced kinship and community as the means of organizing production. (Miller and John, 1992) similarly found that kinship remained a critical component of crew selection in many New England ports, positively impacting the sustainability of both fishing operations and fishing communities. (Knut, 1998) opined that modernization can thus take on a particular flavor when places characterized by more or less close interpersonal ties, egalitarian social networks that are often multiplex and shared identities collide with market relationships that are fleeting impersonal, task oriented and without inherent value.

Junk (2001) understanding the biology and ecology of freshwater systems is critical towards a comprehensive management of flood plain fisheries. Therefore, there is need for a paradigmatic shift, at least in flood plain fisheries to ensure sustainable utilization of the resources. Because of their dynamicity and heterogeneity at both spatial and temporal scales, flood plain fisheries management needs to be approached from a basic understanding of ecosystem ecology of fresh water systems. Junk argues that sustainable development should include the maintenance of the natural flood regime in flood plain systems to ensure sustained fish production. This suggest that external drivers (e.g. flooding), are major drivers of change, which then nullifies assumptions for classical management approaches in flood plain fisheries. (Welcomme, 2001) top down approaches strategies are classical approaches which involve input and output controls which are essentially effort and mesh regulations. According to (Pauly, 1997) classical management approaches are premised on the assumption that fishers are in a "social and financial" position to either comply or implement these measures. However, this is not the case in the developing world, where the challenges of putting food on the table daily are sometimes insurmountable. (Pauly, 1990) stated that the major fisheries management objective in small scale fisheries is to stem the tide of Malthusian over –fishing "Malthusian overfishing" observed that when over fishing occurs poor fishers, without any alternative livelihood strategies, continue fishing even when the resources are severely depleted. In this scenario, women, who would have migrated to urban areas in search of employment, subsidies men fishers through remittances. It can be argued, therefore, that Malthusian overfishing, premised on (Hardin, 1968) "tragedy of the commons" thesis is one of the drivers of fisheries management paradigms in inland fisheries. (Chuenpagde and Jentoft, 2007) highlights that the success of co-management initiatives also depends on pre-implementation stage of the co-management process. Ultimately, co-management is seen as a sustainable response to classical fisheries management failure, which has been predominantly based on top down control measures.

3. CHAPTER THREE: METHODOLOGY

The study area in Jere Local Government Area (Denue, Akawu., Kwayabura & Kida, 2017/Appendix III, Pp. 42), one of the twenty-seven Local Government Areas in Borno State of Nigeria. The Local Government Area was carved out of Maiduguri Metropolitan Council (M.M.C) in 1996 (BSG. 2007). It lies within latitudes 11°40' and 102°5' N and longitudes 13°50' and 12°20' E (MLS, 2008). Within the state, it shares boundaries with Mafa Local Government Area to the east, Maiduguri Metropolitan Council to the north and Konduga Local Government Area to the south. The topography is generally sandy with short grasses and thorny shrubs. Jere Local Government Area has a projected population of 2,93,800 persons with total land area of 868 km^2 with population density of 338.5 per km^2 (NPC, 2016). The majority of the inhabitants are farmers, traders and civil servants. The major ethnic groups are Kanuri and Shuwa Arab. Others include Hausa, Bura, Marghi and Fulani and also many immigrant settlers from within and outside Nigeria (Tijjani, Abubakar, Benisheik and Mustapha, 2010). Part of Maimusari ward was purposively selected out of the 12 wards in the Local Government Area. This is 303 Housing Estate Community, it is located about 812 km North-East of Khaddamari, Head Quarter of Jere Local Government Area and 17.6 km away from Lake Alau "Conventional Lake Chad Basin" and the climate is of arid and desert characteristics: barren dry land with very little rain-fall and vegetation. Practically all rainfalls within three to four months' period from June to September and November to May remain very dry. The annual rainfall recorded in the study area ranges from 500 mm to 700 mm per annum (NMA, 2008). In essence, the study area falls in to the Sudan-Sahelian vegetation classified under the tropical continental climate with rainfall of 250-1000 mm. The temperature regime of the study area is relatively more constant than that of rainfall pattern. The hottest months of the year are March, April and May with mean monthly temperature of 29.5°C, 32.8°C and 34.50°C, respectively (Ezengwu and Okaka, 2001). Civil Service is their major occupation, few compliments it with farming and other small- scale business such as aquaculture, poultry, livestock, sales of local drinks and petty trading, tricycles (Keke-

Napep) transport system, car wash, bread bakery among others (Field-Survey/Interview, 2019/UNCHA, 2019).

The study area has a total population of 2,515 inhabitants (NPC, 2016). The targeted population for this study consists of seventeen households; House No: H04, H20, G03, H28, G22, H14, H11, G31, H05, G07, H43, J28, G12, G02, G01, L26, and L38 associated with aquaculture fish production out of the 303 households in the study area. All the seventeen houses were considered but emphasizes was given to House No: H04 as a result of uninterrupted high scale of production, members of the family skillfulness, experience in aquaculture production system and accessibility for reliable data.

Data for this study was obtained from a primary source. The primary data was obtained through a face to face interview to elicit information from the respondents on aquaculture as a tool in the attainment of community sustainable fisheries development in the study area under the background of economic and cultural consideration.

The population of this study involved all aquaculture fishers in the study area, multi stage sampling technique was employed for selecting the respondents. In the first stage, the houses involved were purposively selected. The second stage was the selection of aquaculture fish farmer in all houses and finally all other members from the selected aquaculture fisher family and other fishers that were involved in aquaculture economic activities in the study area.

Qualitative technique was employed in the analysis of the data and was used to interpret the involvement of fishers in aquaculture and factors militating against the attainment of community sustainable fisheries development in the study area.

4. CHAPTER FOUR: RESULTS AND DISCUSSION

4.1 Economic Considerations

The outputs from the aquaculture farmers were marketed at a ready market in the study area and also served for domestic consumption. Fish production in aquaculture was profitable in the study area, profit of ₦280, 000.00 was realized per harvest "maximum of six months' period" with the used 16×24 Ft size concrete pond (Appendix IV & V, Pp. 43, 44) as the fish matured to table size. The fish farmers were yet to meet maximum requirement to engage in full swing aquaculture production but aquaculture production in the study area was profitable.

This result shows consistency with the findings of Gabriel, Akinrotimi, Bekibele, Omunkwo, and Anyanwu (2007) with steady decline in capture fisheries, aquaculture is readily, veritable tool in the provision of fish eaten all over the continent unlike some other animal products, fish is widely acceptable, its acceptability cut across social, cultural and religious backgrounds.

Fishers in the study area developed interest in aquaculture as a result of skill acquired in a training workshop organized by the federal government of Nigeria but no any additional support services neither from government nor from any other nongovernmental organizations for expansion was provided except from friends, relatives and closed associates e.g. startup capital, assets which was inadequate for further expansion. Majority of the aquaculture fishers in the study area were women (house wives) thus aquaculture was their primary source of income in the study area only few civil servants were engaged in aquaculture practiced to complement their primary source of income.

This result confirmed that of Gabriel, Akinrotimi, Bekibele, Omunkwo, and Anyanwu (2007) to engage in fisheries activities, a start-up capital is needed and fishers should have their initial fund. The three main sources include microcredit, savings, and money lent (often interest-free) by friends or relatives. Most of the women traders (86%) used borrowed money to start-up their business and a little less than half was involved in microcredit (mainly women from Zanziba Town) or had taken a single loan from their husbands or other relative, while the rest borrowed money on a daily basis from a wealthy man in their community.

Aquaculture practice in the study area had enabled other members of the fishers' household acquire knowledge and skills in aquaculture production system through family participatory roles played but another problem asserted was the start-up capital. Although, aquaculture production had led to the emergence of local industries such as smoked fish, fresh fish, dried fish (Appendix VI & VII, Pp. 45, 46), smoked/dried fish retail business, charcoal retail business, mat weaving and traditional smoking kiln retail business.

This result shows inconsistency with the findings of Pauly, (1997) classical management approaches are premised on the assumption that fishers are in a "social and financial" position to either comply or implement aquaculture input and output production measures for the attainment of community sustainable fisheries development. However, this is not the case in the developing world, where the challenges of putting food on the table daily are sometimes insurmountable.

4.2 Cultural Considerations

There was no any cultural constraint to aquaculture practice in the study area, the major hindrance to aquaculture practice in the study area was generally lack of awareness of aquaculture production system as such majority of the people in the study area invested in other forms of business which involved; house rent, keke napep, poultry production, car wash, grinding machine business (engine markade), sales of local drinks, lives stock and crop farming all in small-scale form with the assumption that aquaculture production system requires huge capital investment for operation thus should be handle by government but culturally, right from history there was no any cultural taboo attached to fish farming in the study area.

This result confirmed that of Patricia and Julia (2007) characteristics of fishing communities range from containers of cultural traditions to sites of resistance, from place of management to space of global connection.

Culturally, it was the traditional authorities that was empowered on the decision of the society but this was not applied in the activities of aquaculture in the study area; the nomination of trainees to participate in aquaculture skill acquisition training was selected based on qualification, indigenization, gender and geographical location regardless of the long tradition the state had that has been deeply rooted in traditional jurisprudence of the state.

This result shows consistency with the findings of Knut (1998) modernization can thus take on a particular flavor when places characterized by more or less close interpersonal ties, egalitarian social networks that are often multiplex and shared identities, collide with market relationships that are fleeting impersonal, task oriented and without inherent value.

Fish has been culturally and nutritionally important source of food among the diet of the people in the study area and the people have attached very high value for fish product inconsiderate of easy accessibility of meat than fish in the study area but yet people in the study area prepared fish to meat.

In order to ensure rational utilization of fisheries resources; culturally, the traditional authorities were responsible for the allocation of land for fishing in fishing communities under the jurisdiction of the village head (Bulama) but this was not obtainable in the study area as the system of fish production in the study area has been home based aquaculture production system. Culturally, there was a believed that for a woman to mingle with opposite sex in open water for fishing was a taboo; this cultural notion had negatively influenced the level of participation of women in fishing activities.

The reverse was the case; women fishers practiced aquaculture in households had positively influenced the level of participation of women in the study area. Culturally, land for fishing was allocated by traditional authorities to fishers in fishing communities had not constitutes constraints to fishing activities. More so there was no any land use act and or legislation applied to aquaculture as constraint to aquaculture by modern authorities in the study area (Appendix I, II & IV, Pp. 40, 41, 43). Majority of the aquaculture fish farmers in the study area used the cultural traditional method of harvesting, sun drying and smoking in fish processing (Appendix V & VII, Pp. 44, 46).

This result shows inconsistency with the findings of Touray (1997) several factors seem to contribute to the inferior position and suppressed rights and privileges of women in society. Tradition and culture, social and religious norms and values, political and economic factors all contribute to the lowering of the status of women in society and contribute to the form of dominance to which women are subjected to. They tend to separate roles and responsibilities for men and women in society and hence bar women from sharing equal rights, opportunities, and privileges and prevent them from effectively participating equally with their men counterparts in the sustainable development of the community. These have also given rise to restrictions in the levels and degrees of participation by women members of the community and they suppress the ideas and potentials of women.

4.3 Integrating Aquaculture and Sustainable Development

Findings revealed that majority of the aquaculture fishers in the study area were women and engaged only on concrete ponds and collapsible pond/tank aquaculture production system. Aquaculture was profitable in the study area, the fishers prioritized aquaculture production system rather than any other form of fish production, it had contributed positively to fishers' livelihood and enhances sustainable fisheries development.

Aquaculture was not in any way contradictory to the principle of women in fishing neither culturally nor religiously in the study area. The under-utilized resources as a result of cultural constraints faced by women that hindered women participation in the fisheries sector had been better off consequently production gap was recouped and had multiplier effect on economic activities in the study area (Appendix IV & VII, Pp. 43, 46). This had cumulated positively to sustainable fisheries development in the study area. The interdependency among fishers in the study area had promoted better management of limited resources both technically and otherwise, more so created better access to aquaculture resources both input and output in a facilitated manner had contributed to community sustainable fisheries development in the study area.

This result confirmed that of Chuenpagde and Jentoft (2007) the success of co-management initiatives also depends on pre-implementation stage of the co-management process. Ultimately, co-management is seen as a sustainable response to classical fisheries management failure, which has been predominantly based on top down control measures.

Support system from government and other non-governmental organization to the aquaculture fishers in the study area was poor had contributed negatively in sustaining the aquaculture fisheries in the study area and also non-existence of the application of land use act and aquaculture legislation in the sub-sector in the study area was expressed as advantageous in the short run by the fishers in the study area and contemplated the long run resultant effect would be negative to community sustainable fisheries development in the study area (Appendix I & II, Pp. 40, 41).

This result confirmed that of FAO (2013) Nigeria has no legislation regarding aquaculture at national level. It is not directly mentioned in the Sea Fisheries Degree (1971, 1992) but the ministry of fisheries is to determine whether the setting up of enclosures, pens, and cages should be subjected to license fees.

5. CHAPTER FIVE: CONCLUSION

Aquaculture has great potentials for sustaining fish production in 303 housing estate community of Jere Local Government Area, Borno state of Nigeria if well harnessed. The need to give more emphasis to aquaculture production system considering the fact that catch from inland fisheries have been overexploited with the attendant problem of insecurity. In order to enhance fish production to ensure sustainable fisheries development aquaculture opportunities should be properly harnessed.

RECOMMENDATIONS

The following recommendations are made:

1. Government and other non-governmental organizations should be duly informed to inculcate the concept of aquaculture fisheries in the mind of members of the fishing communities in relation to the role aquaculture plays in the attainment community of sustainable fisheries development.
2. Community fishers (households) that have undergone training should be empowered with adequate aquaculture skills, adequate capital and other material requirement to embark on aquaculture practice in a sustainable manner in the study area; House No: H04, H20, G03, H28, G22, L38, H14, H11, G31, H05, G07, H43, J28, G12, G02, G01 and L26. Alternatively, fish farming entrepreneurship for youths in Borno state of Nigeria should be organized with the theme; "Achieving Sustainable Community Fisheries Development through Youths Empowerment"
3. Community level aquaculture cooperative societies and associations should be set up to relate jointly with government and other non-governmental organizations to identify the need of the community fishers through equal community participatory role to initiate and implement a project targeted toward ensuring community sustainable fisheries development.

4. I don't care attitudes, corruptions; unequal treatment, selfish interest and all other obstacles considered repugnant to the attainment of sustainable fisheries development at all levels of project initiators, implementers and or executors should be eliminated.
5. Appropriate aquaculture land use act and aquaculture legislation framework in consonance to the attainment of community sustainable development should be put in place.

The study is of very importance because it enables realization of the economic viability of aquaculture fish farming as a tool for food security and sustainable development and provides ground work as a basis for the conduct of pre-feasibility and feasibility studies necessary for the establishment of concrete pond fish farm and ascertain it workability through participatory aquaculture in the attainment of sustainable fisheries development.

REFERENCES

Abbas A. M, Damisa M., and Ahmed B. (2015) Economics of Fish Farming in Ondo State, Nigeria: A Profit Function Approach. Journal of Global Agriculture and Ecology. 2015; 4(2): 93-98. https://www.ikpress.org/index,php/JOKBAE/article/view/2756

Abdalla P. R. and Sumaila U. R. (2007) A Historical account of Brazilian Public Policy on Fisheries Subsidies. Marine policy, 31(4), 444-450, doi: 10.1016/j. marPol.2006.12.002. https://www.sciencedirect.com/science/article/pii/S0308597X07000098

Barth F. (1992) Towards Greater Naturalism in Conceptualizing Societies. In Conceptualizing Society. Adam Kuyper. London: Routledge. 1992; 1733.

BSG. (2007) Borno State Government (BSG) Official Diary of Ministry of Information, Home Affairs, Maiduguri, Nigeria. 2007; 5-7.

Chuenpagde R, and Jentoft S. (2007) Step Zero for Fisheries Co-Management: What Precedes Implementation? Marine Policy. 2007; 31: 657-668.

Cunning L. (2010) Assessing the Contribution of Aquaculture to Food Security: A Survey of Methodologies. FAO Fisheries Circular No.1010. Rome, Italy: FAO, 2010. http://www.fao.org/3/a-i1010e.pdf

Denue, BA., Akawu, CB., Kwayabura, SA., Kida I, (2017) Low Case Fatality During 2017 Cholera Outbreak in Borno State, North Eastern Nigeria. Annals of African Medicine, 2018;17::203-9. Published by Wolters Kluwer – Medknow. DOI: 10.4103/aam.aam-66-17

Ezengwu SM, and Okaka CE. (2001) Some Health Hazards Associated with Intensive Fishing Activities at the Alau Dam, Borno State, Nigeria. J Arid Zone Fish. 2001; 1: 33-38. http://www.annalsafrmed.org

FAO (2006) Food and Agricultural Organization of the United Nations. Contribution of Fisheries to National Economies in West and Central Africa – Policies to increase the Wealth generated by Small Scale Fisheries. New Direction in Fisheries – A Series of Policy Briefs on Development Issues Rome. 2006; 12. http://www.fao.org/3/a-a0785e.pdf

FAO (2007) Food and Agricultural Organization of the United Nation, Fisheries Development in Nigeria, the Current Challenges: Paper Presented by the Hon. Minister of State for Agriculture and Rural Development, Fisheries Society of Nigeria (FISON) Conference 2007, Lagos State. 2007; 23.

FAO (2013) Food and Agricultural Organization of the United Nation, Fisheries Development in Nigeria, the Current Challenges: Paper Presented by the Hon. Minister of State for Agriculture and Rural Development, Fisheries Society of Nigeria (FISON) Conference 2007, Lagos State, Nigeria, 2013.

FAO (2014) Food and Agricultural Organization of the United Nations (FAO). The Fisheries Sector Contribution to Poverty Alleviation and Food Security. The State of World Fisheries and Aquaculture: Opportunities and Challenges. Food and Agricultural Organization, Rome, Italy. 2014. http://www.fao.org/3/a-i3727e.pdf

Fischer CS. (1975) The Study of Urban Community and Personality. Annual Review of Sociology. 1975; 1: 67-89.

Frocklin S, De la Torre-Castro M, Lindström L, Jiddawi NS, and Msuya FE. (2012) Mariculture as a Development Project in Zanzibar, East Africa: A price too high to pay? Aquaculture. 2012; 357:30–39.

Gabriel UU, Akinrotimi OA, Bekibele DO, Omunkwo DN., and Anyanwu PE (2007) Locally Produced Fish Feed: Potentials for Aquaculture Development in Sub-Saharan Africa. African Journal of Agriculture Research. 2007; 2: 287-295.

Garcia S. M, and Rosenberg A. A. (2010) Food Security and Marine Capture Fisheries: Characteristics, Trends, Drivers and Future Perspectives. Philosophical Transactions of the Royal Society B: Biological Sciences,365(1554), 2869-2880.doi:10.1098/rstb:2010.0162. https://royalsocietypublishing.org/doi//10.1098/rstb.2010.0162

Hardin G. (1968) The Tragedy of the commons. Science. 1968; 162:12431248.

Haylor G, and Bland S. (2001) Integrating Aquaculture in to Rural Development in Coastal and Inland Areas, Technical Proceedings of the Conference on Aquaculture in the Third Millennium, Bangkok, Thailand, 20th – 25th February, 2001.

Inoni OE. (2001) Allocative Efficiency in Pond Fish Production in Delta State, Nigeria: A production function approach. Agriculture Tropical Et Subtropical. 2001; 40: 144-152.

Junk WJ. (2001) Sustainable Use of the Amazon River Flood Plain: Problems and Possibilities. Aquatic Ecosystem Health and Management. 2001; 3: 225-233.

Knut H. (1998) Community, State and Market on the North Atlantic Rim: Challenges to Modernity in the Fisheries. Toronto: University of Toronto Press. 1998.

Kurlansky M. (1997) Cod: A Biology of the Fish that change the World. New York: Penguin. 1997.

Lipuma E. (1992) Social Identity and the European Community. Maritime Anthropological Studies. 1992; 5: 46-73.

Massey D. (1993) Power Geometry and a progressive Sense of Place in Mapping the Futures: Local Cultures, Global Change. John Bird Curtis, Tim Putnam, George Robertson, and Lisa Thickener. 1993; 59-69.

Miller M, and John VM. (1992) Boats Don't Fish, People Do. Some Ethnographic Notes on the Federal Management of Fisheries in Gloucester. In Essays on Social and Cultural Aspects of New England Fisheries. Report to the National Science Foundation. James Acheson, ed 1992; 2: 27-53.

MLS. (2008) Ministry of Land and Survey, Maiduguri, Borno State, Nigeria, 2008 Office Memo File. 2008; 4: 55-58.

NMA (2008) Nigerian Metrological Agency, 2008 Annual Report. Office Memo File. 2008.

Nnodim O. (2014) Nigerian Spends ₦125 Billion Annually on Fish Importation. 2014. The Punch Newspaper. https://punching.com/nigeria-spends-n125-billion-annually-on-fish-importation/

NPC (2016). National Population Commission of Nigeria, 2006 Census. 2016.

Okoye F. C, and Ayanda J. O. (2004) Economic of Commercial Fish Farming in Earthen Ponds in Nigeria: Privatization of Fish Farms, National Institute for Freshwater Fisheries Research P.M.B. 6006, New Bussa, Niger State. 2004.

Olufayo MO. (2012) Gender Roles of Women in Agriculture and Food Security in Nigeria, Department of Fisheries and Aquaculture Technology, Federal University of Technology Akure, Nigeria, IIFET 2012 Tanzania Proceedings. 2012.

Patricia MC, and Julia O. (2007) Defining Fishing Communities: issues in theory and practice. NEPA BULLETING. 2007; 28: 27-42.

Pauly D. (1990) On Malthusian Overfishing. NAGA. 1990; 13: 3-4.

Pauly D. (1997) Small Scale Fisheries in the tropics: Marginality, Marginalization for Fisheries Management. 1997.

Raji A, and Omoyemi B. (2005) Resource Utilization Practices and Sustainable Development at the Lake Chad Basin, Federal College of Freshwater Fisheries Technology, P.M.B 1060, Maiduguri, Borno State. 2005.

Rogers SA. (1991) Shaping Modern Times in Rural France: The Transformation and Reproduction of an Aveyronnais Community. Princeton University Press. 1991.

Serge MG. (2007) The FAO definition of Sustainable Development and the Code of Conduct for Responsible Fisheries, An Analysis of the Related Principles, Criteria and Indicators. 1998.

Stella BW. (2000) Economic Potentials of Women in Small-Scale Fisheries in West Africa. 2000.

Tijjani BA, Abubakar M, Benisheik KM, and Mustapha AB. (2010) Resource Use Efficiency in Rice Production in Jere Local Government Area of Borno State, Nigeria. Nigerian Journal of Basic and Applied Science. 2010; 18: 27-34.

Touray I. (1997) Gender issue in the Fisheries Sector and Effective Participation. Programme for Integrated Development of Artisanal Fisheries in West Africa (IDAF Programme). Technical Report, 1997.

UNCHA (2019). United Nation for the Coordination of Humanitarian Affairs. 2019.

Welcomme RL. (2001) Inland Fisheries: Conservation and Management Black Wells Oxford. 2001; 350.

APPENDIX I:

Disposed Waste Water within Household Compound in the Study Area

Source: Field-Survey/Interview, 2019

APPENDIX II:

Disposed Waste Water from Fisherwoman Household in the Study Area

Source: Field-Survey/Interview, 2019

APPENDIX III:

Study Area Map Borno State Showing 27 Local Government Areas

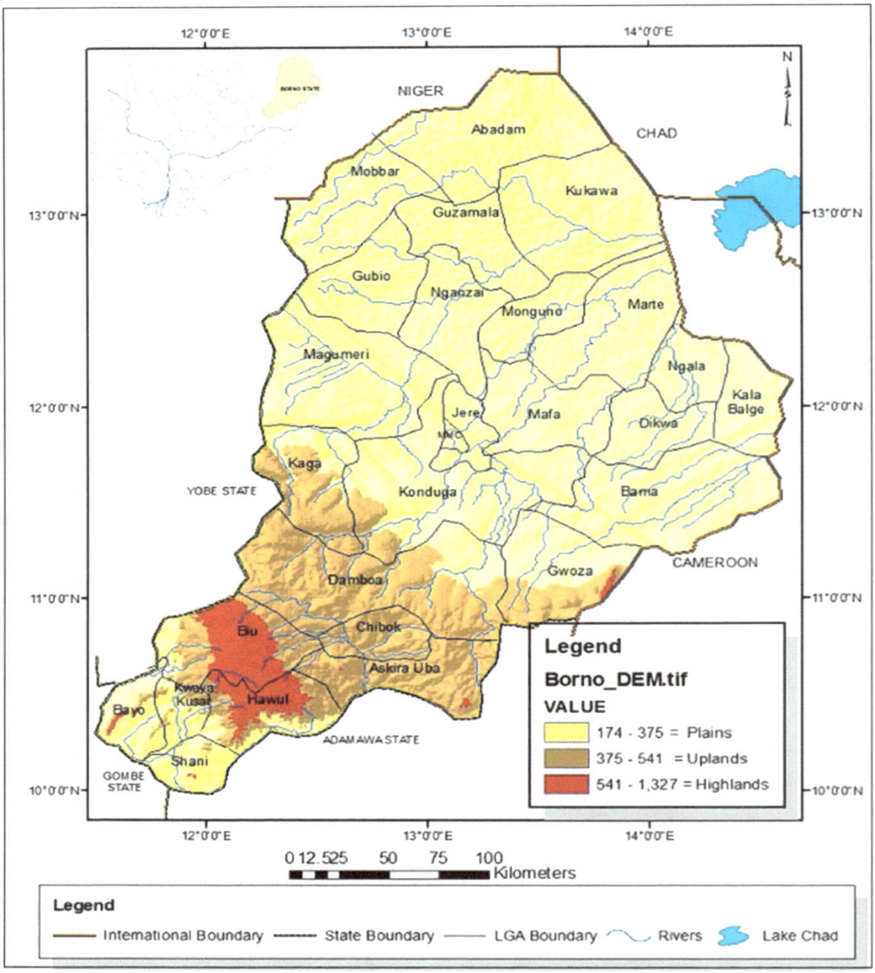

Source: Denue, Akawu, Kwayabura, Kida (2017)

APPENDIX IV:

Household Concrete Fish Pond (16 × 24 Ft) Fisherwoman replacing Water in the Study Area

Source: Field-Survey/Interview, 2019

APPENDIX V:

Fresh Fish Dried for Household Domestic Consumption and Fish Sold at ₦900.00 per Kilogram in the Study Area

Source: Field-Survey/Interview, 2019

APPENDIX VI:

Smoked Fish Retail Businessman in the Study Area

Source: Field-Survey/Interview, 2019

APPENDIX VII:

Semi-skilled Fisher lady in Fish Processing Stage in the Study Area

Source: Field-Survey/Interview, 2025

www.ingramcontent.com/pod-product-compliance
Lightning Source LLC
LaVergne TN
LVHW051131210926
848482LV00001B/65